Abdelhafid Mimouni

Nitrates and Nitrites : Risks and Answers

Abdelhafid Mimouni

Nitrates and Nitrites : Risks and Answers

ScienciaScripts

Imprint

Any brand names and product names mentioned in this book are subject to trademark, brand or patent protection and are trademarks or registered trademarks of their respective holders. The use of brand names, product names, common names, trade names, product descriptions etc. even without a particular marking in this work is in no way to be construed to mean that such names may be regarded as unrestricted in respect of trademark and brand protection legislation and could thus be used by anyone.

Cover image: www.ingimage.com

This book is a translation from the original published under ISBN 978-620-6-72235-9.

Publisher:
Sciencia Scripts
is a trademark of
Dodo Books Indian Ocean Ltd. and OmniScriptum S.R.L publishing group

120 High Road, East Finchley, London, N2 9ED, United Kingdom
Str. Armeneasca 28/1, office 1, Chisinau MD-2012, Republic of Moldova, Europe
Printed at: see last page
ISBN: 978-620-8-12413-7

Nitrates and Nitrites : Risks and Responses

Author : Dr. Abdelhafid Mimouni

An independent researcher in bioinorganic chemistry, Dr Mimouni is an expert in macromolecular synthesis and characterisation. He obtained his PhD in Chemistry from the University of Paris XII in 1997, after a Diplôme des Études Approfondies in Bioinorganic Systems from the University of Paris XI in 1993, where he also obtained his Licence and Maîtrise in Chemistry.

Summary: Nitrates and nitrites are chemical compounds often used in food preservation, particularly in processed meats such as sausages and bacon. They help prevent the growth of bacteria, including *Clostridium botulinum*, a dangerous microbe that can cause severe food poisoning. Nitrites react with myoglobin, a protein found in muscles, to give processed meats their attractive pink colour. However, when consumed in excess, nitrites can turn into potentially harmful substances. It is therefore crucial to regulate their use to protect health while maintaining food quality.

Plan :

1. **Chemical structure of nitrates and nitrites**

   o **Composition and chemical properties** Nitrates (NO_3^-) and nitrites (NO_2^-) are inorganic anions derived from nitrogen, playing crucial roles in various biological and industrial processes. Nitrate, with its formula NO_3^-, is a stable and relatively inert ion. It is widely used in agriculture in the form of fertilisers such as potassium nitrate (KNO_3) and sodium nitrate ($NaNO_3$). In the food industry, nitrate is mainly used to preserve meat. In comparison, nitrite, whose formula is NO_2^-, is much more reactive. Used primarily to inhibit microbial growth in processed foods, it also helps stabilise the colour of meat products. The greater chemical reactivity of nitrites, compared with nitrates, makes them of crucial importance in food safety, although this reactivity also gives rise to health concerns due to the potential formation of harmful compounds.

   o **Comparison between nitrates and nitrites** Nitrates, as less reactive ions, are mainly used as preservatives in processed meats and as agricultural fertilisers. Nitrites, which are more reactive, are involved in chemical processes that can lead to the formation of nitrosamines when they react with amines in the stomach. Nitrosamines are potentially carcinogenic compounds, which raises concerns about the excessive consumption of nitrites. This difference in reactivity justifies the distinct uses of these two ions and the risks associated with each.

2. **Conversion mechanisms**

   o **Transformation of nitrates into nitrites in the body** When ingested, nitrates undergo a reduction into nitrites, a process that takes place mainly in the oral cavity and intestines, facilitated by specific bacteria in the microbiota. In the mouth, nitrate-reducing bacteria, particularly those belonging to the Neisseriaceae and Veillonellaceae families, convert nitrates into nitrites. This process continues in the intestines, where bacteria such as *Escherichia coli* and *Lactobacillus* also contribute to this conversion. This mechanism is essential to the metabolism of nitrates and nitrites in the human body, playing a role in the regulation of blood pressure and other physiological functions.

   o **Role of nitrate-reducing bacteria** Nitrate-reducing bacteria are crucial for the conversion of nitrates into nitrites. In the oral cavity, these bacteria, found mainly on the tongue and in saliva, are responsible for reducing nitrates. In the intestinal tract, bacteria such as *Escherichia* *coli* and *Lactobacillus* continue this conversion. This bacterial process is not only vital for digestion, but also for the effectiveness of nitrites in food preservation processes. Variability in bacterial populations and physiological conditions can influence the rate of conversion and therefore the effectiveness of nitrates and nitrites.

3. **Industrial use in food**

   - **Use of nitrates in processed meats** Nitrates are commonly used in the processing of meats such as ham, sausages and bacon. Their main role is to prevent the growth of *Clostridium botulinum*, a bacterium responsible for botulism, a potentially fatal food poisoning. In the presence of nitrates, conditions become unfavourable for the growth of this bacterium, making meat products even safer. The conversion of nitrates to nitrites during food processing is also essential for maintaining the characteristic pink colour of processed meats.

   - **Other food applications** In addition to their use in meat, nitrates and nitrites are present in other food products. For example, nitrates are sometimes added to vegetables such as beetroot and spinach to prolong their shelf life. Because of their ability to fix colour, nitrites are widely used in charcuterie to improve the appearance of meat products.

These compounds also play an important antimicrobial role, which is crucial to the safety of processed foods. However, the use of these compounds in the food industry is strictly regulated because of the potential health risks.

Chapter 2: Microbiology of Nitrates and Nitrites

1. Nitrate-reducing bacteria

a. Presence in the oral cavity: Neisseriaceae and Veillonellaceae

Nitrate-reducing bacteria present in the oral cavity play a crucial role in the conversion of nitrates into nitrites, a process vital to human physiology. Among these bacteria, the Neisseriaceae and Veillonellaceae families stand out for their ability to convert nitrates into nitrites using specific enzymes. Located on the tongue and in saliva, these bacteria initiate the reduction of nitrates at the start of the digestive process, thereby promoting the activation of nitrites, which play a key role in regulating a number of biological functions.

b. Presence in the intestinal tract: Escherichia coli, Enterobacter, and Lactobacillus

In the intestines, nitrate reduction continues under the action of other bacteria such as Escherichia coli, Enterobacter and Lactobacillus. These micro-organisms, which form an integral part of the intestinal microbiota, are responsible for converting nitrates into nitrites, thereby influencing the dynamics of these compounds in the digestive system. Their role is essential not only for the metabolism of ingested nitrates, but also for overall digestive health, since these bacteria modulate the microbiotic balance and have an impact on the bioactivity of nitrites.

2. Antimicrobial effect of nitrites in foodstuffs

a. Inhibition of Clostridium botulinum

One of the most important applications of nitrites in the food industry is their ability to inhibit the growth of Clostridium botulinum, a bacterium that produces a deadly toxin in anaerobic environments, such as those found in poorly preserved meat. Nitrites interact with the myoglobin present in meat muscles, creating an unfavourable environment for this bacterium and thus reducing the risk of botulism. This antimicrobial property makes nitrites an essential preservative in food safety.

b. Effects on other food pathogens

In addition to their action against Clostridium botulinum, nitrites have antimicrobial properties against a variety of other food pathogens, including certain strains of Salmonella and Escherichia coli. The effectiveness of nitrites in inhibiting the growth of these bacteria contributes to the prevention of food poisoning, thereby enhancing the safety of processed meat products. This extensive antimicrobial action explains why nitrites are commonly used in food preservation, providing protection against a wide range of pathogens.

3. Impact of nitrites on the colour and taste of meat

a. Reaction of nitrites with myoglobin

Nitrites play a fundamental role in changing the colour of processed meats. When they react with myoglobin, they form nitrosomyoglobin, a complex that gives meat its characteristic pink colour, often associated with freshness and quality. This chemical reaction is essential for the visual appeal of meat products, an important visual quality for consumers.

b. Role in maintaining meat quality

In addition to their impact on colour, nitrites also help to preserve the organoleptic quality of meat by affecting taste and texture. They prevent the oxidation of lipids, delaying rancidity, and inhibit the growth of micro-organisms responsible for spoilage. As a result, nitrites extend the shelf life of processed meats while maintaining their taste and texture qualities, meeting consumer demands for freshness and flavour.

Chapter 3: Impact on human health

1. Formation of methaemoglobin

a. Process of methaemoglobin formation

Methaemoglobin is an altered form of haemoglobin in which ferrous iron (Fe^{2+}) is oxidised to ferric iron (Fe^{3+}), a transformation catalysed by nitrites when ingested. This process begins in the digestive system, where nitrites, after absorption, enter the bloodstream. Once in the blood, they react with haemoglobin, leading to the formation of methaemoglobin. This phenomenon is particularly favoured by the acidic conditions present in the stomach and intestinal tract, creating an environment conducive to oxidation.

b. Clinical consequences: methaemoglobinaemia

The accumulation of methaemoglobin in the blood leads to a condition known as methaemoglobinaemia. This condition reduces the blood's ability to carry oxygen, which can lead to symptoms such as headaches, dizziness, breathing difficulties and, in severe cases, cyanosis, a blue discolouration of the skin caused by hypoxia. Infants, because of their immature enzyme systems, and people with specific enzyme deficiencies, are particularly at risk of developing this condition, underlining the importance of monitoring exposure to nitrites.

2. Nitrosamines: carcinogenic compounds

a. Formation of nitrosamines in the stomach

Nitrosamines, potentially carcinogenic compounds, are formed when nitrites react with secondary or tertiary amines in an acidic environment such as the stomach. This chemical process is amplified by the acidic pH of the stomach and the presence of amino precursors derived from the digestion of food proteins. Nitrosamines are known to have carcinogenic properties, making them a major public health concern, particularly because of their potential to cause cell mutations.

b. Risks associated with high nitrite consumption

Excessive consumption of nitrites is linked to increased health risks, particularly the development of cancer. Nitrosamines formed in the digestive system can interact with DNA, causing genetic mutations and increasing the risk of cancer, particularly of the gastrointestinal tract. Epidemiological studies have established a link between a diet rich in processed meats containing nitrites and an increased risk of certain cancers, highlighting the importance of monitoring nitrite intake.

3. Epidemiological and toxicological studies

a. Analysis of health risks

Epidemiological studies have explored the long-term effects of nitrite and nitrate consumption on human health, revealing a correlation between high consumption and the incidence of certain types of cancer, particularly those affecting the digestive tract. At the same time, toxicological studies have shed light on the mechanisms by which nitrites and nitrosamines induce cellular damage, contributing to the development of chronic diseases. This research reinforces the need to regulate nitrite levels in food and encourage safe eating practices to minimise health risks.

b. Recommendations from the health authorities

Public health authorities, such as the World Health Organisation (WHO) and the European Food Safety Authority (EFSA), are recommending a reduction in the consumption of nitrites and nitrates to reduce the risks associated with their use. These bodies have set maximum limits for nitrite levels in food and are promoting safer alternatives for food preservation. These recommendations aim to protect consumer health while allowing the controlled and safe use of nitrates and nitrites in the food industry.

Chapter 4: Regulations and food safety

1. **Standards and concentration limits** a. **International regulations (EU, FDA, etc.)**

Nitrates and nitrites in food are strictly regulated to protect public health, with standards varying from region to region but following similar principles to ensure food safety:

- **European Union (EU)**: Regulation (EC) No 1333/2008 governs the use of nitrates and nitrites by setting maximum limits for these compounds in various types of food. For example, for processed meats such as bacon and sausages, the maximum concentration of nitrites is often limited to 150 mg/kg. These limits minimise the risk of nitrosamine formation, while allowing these additives to be used for preserving and colouring food products.

- **Food and Drug Administration (FDA)**: In the United States, the FDA regulates nitrates and nitrites as food additives in the Code of Federal Regulations (CFR) Title 21. Maximum permitted limits are similar to those in the EU, with a general limit of 200 ppm for nitrites in processed meats, aimed at ensuring safe levels for human consumption.

b. **Maximum permitted limits in foods**

Maximum limits for nitrates and nitrites are set according to the type of food and their role in preservation or colouring:

- **Processed meats** : Nitrite levels generally vary from 50 to 200 mg/kg depending on product specifications and national regulations, to ensure consumer safety while preserving food.

- **Dairy products and other foods**: Limits are often stricter, reduced to minimum levels to reflect the different uses and preservation needs in these products.

2. **Alternatives to nitrates and nitrites a. Natural substitutes and new technologies**

In response to concerns about nitrates and nitrites, several alternatives have been developed to maintain food quality:

- **Natural substitutes**: Alternatives such as celery extract and beetroot powder are used as natural sources of nitrates, which turn into nitrites during food processing. These substitutes are popular in 'no added nitrate' products for their more natural perception.

- **New technologies**: Ozonation and the use of natural antioxidants such as vitamin C (ascorbic acid) are being explored to reduce or eliminate the need for nitrites, while maintaining food safety and minimising the associated risks.

b. **Advantages and disadvantages of the alternatives**

- **Advantages** :
  - **Reducing health risks**: Natural substitutes can reduce the formation of nitrosamines and reduce the risks associated with nitrites.
  - **Growing acceptability**: Natural products without synthetic additives are becoming increasingly popular with consumers.
- **Disadvantages** :
  - **Cost**: Natural alternatives and new technologies can be more expensive, increasing the price of finished products.
  - **Variable effectiveness**: Some substitutes may not offer the same levels of preservation or colour, affecting the quality of food products.

3. **Industry practices and liability a. Food industry practices**

The food industry plays a crucial role in implementing

regulations to ensure that nitrate and nitrite levels comply with safety standards:

- o **Quality control**: Manufacturers need to monitor and regulate nitrate and nitrite levels throughout the production process. Regular checks and testing are essential to maintain product safety.

- o **Labelling**: Products containing nitrates and nitrites must be correctly labelled to inform consumers of their presence. Transparent labelling helps consumers to make informed choices.

b. **Impact of industrial choices on public health**

Food industry decisions have a direct impact on public health:

- o **Risk reduction**: Adopting alternatives to nitrates and nitrites can reduce the risk of nitrosamine formation and improve food safety.

- o **Raising awareness**: Transparent labelling and educating consumers about the associated risks encourage more informed decisions on the part of consumers.

Chapter 5: Case studies and current debates

1. **Case studies**

a. **Examples of food products and their nitrate/nitrite content**

Nitrates and nitrites are widely used in the food industry for their preservative and colouring properties. Here are some examples of food products containing these compounds:

- **Charcuterie**: Charcuterie products such as bacon, ham and sausages use nitrites to prevent bacterial growth and improve colour and flavour. For example, bacon can contain up to 150 mg/kg of nitrite. Nitrites prevent the proliferation of pathogenic bacteria such as *Clostridium botulinum*, while giving the product its characteristic pink colour.
- **Processed meats** : Hot dogs, pâtés and other processed meats contain significant levels of nitrites, generally between 50 and 150 mg/kg. These levels are regulated to maintain colour and flavour while complying with food safety standards.
- **Bakery products**: Although less common, nitrates can be used in certain bakery products such as sourdough bread, where they act as yeast. In these products, nitrates are

converted to nitrites during the baking process, contributing to the texture and preservation of the product.

Nitrate and nitrite levels in these products are regularly monitored to comply with local and international regulations, guaranteeing consumer safety while ensuring effective preservation.

b. Analysis of botulism cases linked to poor preservation

Botulism is a serious food poisoning caused by the toxin produced by *Clostridium botulinum*. Errors in food storage can lead to outbreaks of botulism, as illustrated by the following cases:

- **Historical example**: In 2018, an outbreak of botulism was reported in poorly preserved canned meat products in a small establishment. A failure of proper cooking procedures allowed the survival of *C. botulinum*, which produced the toxin in the canned goods. The offending products contained high levels of nitrates and nitrites, but poor temperature management and sterilisation practices were the main causes of the outbreak. This situation has highlighted the importance of proper cooking and storage, even when high levels of nitrites are present.

o **Recent case**: In 2020, an outbreak of botulism was linked to home-made sausages that did not comply with food safety standards. The products were insufficiently treated to eliminate the bacterium, and nitrite levels were present, but were not sufficient to prevent the development of the toxin due to poor storage management. This case revealed that the presence of nitrites does not guarantee food safety if storage conditions are not respected.

These cases highlight the need for strict compliance with food safety standards to prevent the growth of pathogenic bacteria and ensure that nitrate and nitrite levels are correctly regulated and used.

2. **Current debates on the use of nitrites**

a. **Industry outlook versus consumer concerns**

o **Industry outlook**: The food industry justifies the use of nitrates and nitrites because of their effectiveness as preservatives and colouring agents. These compounds extend the shelf life of products, improve their appearance and help maintain food safety standards by inhibiting the growth of pathogenic bacteria. Industry arguments include the need to guarantee food safety and the ability to keep prices affordable for consumers. Nitrites also play a crucial role in preventing the formation of toxins in meat products.

o **Consumer concerns**: Consumers are expressing growing concern about the health risks associated with nitrates and nitrites, in particular the formation of nitrosamines, potentially carcinogenic compounds. Conscious consumer movements are pushing for products with no added nitrates, and concerns are reinforced by studies linking high nitrite consumption to various health problems, including certain types of cancer. Awareness campaigns are highlighting the potential health risks associated with nitrites and promoting safer alternatives.

b. **Initiatives to reduce or replace nitrates/nitrites**

o **Natural substitutes**: In response to concerns about nitrates and nitrites, a number of alternatives have been developed to replace these additives while maintaining food quality. Celery extract and beetroot powder, rich in natural nitrates, are used as substitutes. These natural nitrates are converted to nitrites during food processing, offering an option perceived as more natural and less risky by consumers.

- **Alternative preservation technologies**: Innovative technologies, such as ozonation and the use of natural antioxidants like vitamin C (ascorbic acid) and rosemary extracts, are being explored to reduce or eliminate the need for nitrites. These methods aim to preserve food safety while reducing the risks associated with nitrites, offering viable alternatives to traditional practices.

- **Regulatory initiatives**: Some countries and health organisations are recommending reducing maximum permitted levels of nitrites in food or promoting production practices that minimise the risk of nitrosamine formation. These recommendations may include stricter guidelines for additive levels and advice for consumers on a balanced diet, aimed at balancing conservation benefits with public health concerns.

Chapter 6: Strategies for reducing or eliminating Nitrites from the Body

1. **Dietary approaches and supplements**

a. **Low nitrate diets**

Diet plays a fundamental role in managing nitrate intake, influencing nitrite formation in the body. Adopting a low-nitrate diet can contribute to a significant reduction in nitrite levels. Here are some specific strategies:

- o **Low-nitrate diets**: Avoiding nitrate-rich foods, such as certain leafy vegetables (spinach, lettuce), beetroot, and processed meat products, can help reduce exposure to nitrates. For example, a diet based on fresh, unprocessed foods limits nitrate intake, which can reduce its conversion to nitrite. Mediterranean diets, which emphasise the consumption of fresh vegetables, fruit, whole grains and oilseeds, are often low in nitrates and can serve as a model for a balanced diet.

- o **Practical examples**: Organic produce, which tends to contain less nitrates due to the absence of chemical fertilisers, can also reduce overall nitrate intake. Studies have shown that organic fruit and vegetables generally have lower nitrate levels than conventional produce. Preferring to prepare fresh food rather than eating prepackaged products containing added nitrates is also recommended. Cooking vegetables at high temperatures can reduce nitrate levels in food.

b. **Food supplements and antioxidants**

Food supplements and antioxidants can play an important role in neutralising the harmful effects of nitrites. Here are a few options:

- o **Vitamin C**: Vitamin C is a powerful antioxidant that helps neutralise nitrites by preventing their conversion into nitrosamines, potentially carcinogenic compounds. It is commonly added to food products as a preservative and can be taken as a supplement to counter the effects of nitrites. Studies have shown that high doses of vitamin C can reduce the levels of nitrosamines in the body.
- o **Polyphenols**: Polyphenols, found in foods such as berries, green tea and cocoa, have antioxidant properties that help neutralise nitrites and protect cells from oxidative damage. Polyphenol supplements, such as those based on green tea extract or berries, can also be considered to reduce the effects of nitrites in the body. Research suggests that

polyphenols may reduce the absorption of nitrites and their transformation into nitrosamines.

2. **Modulation of the microbiota**

a. **Prebiotics and targeted probiotics**

Modulating the gut microbiota is a promising strategy for managing nitrite levels. Here's how prebiotics and probiotics can help:

- **Prebiotics**: Prebiotics are dietary fibres that promote the growth of beneficial bacteria in the intestinal microbiota. A balanced microbiota can reduce nitrite production by minimising the presence of nitrate-reducing bacteria. Foods such as bananas, onions, asparagus and chicory root are rich in prebiotics and promote healthy intestinal flora.
- **Probiotics**: Probiotics can also help maintain a healthy balance of intestinal bacteria. Some probiotic strains, such as those found in yoghurts and probiotic supplements, are able to reduce the activity of nitrate-reducing bacteria, thereby limiting nitrite production. Research has shown that specific strains, such as *Lactobacillus* and *Bifidobacterium*, can have an inhibitory effect on the conversion of nitrate to nitrite.

b. **Specific inhibitors of metabolic pathways**

Compounds capable of inhibiting the enzymes responsible for reducing nitrates to nitrites are currently being explored. These inhibitors could offer targeted solutions:

- **Inhibiting compounds**: Research is underway into compounds such as flavonoids, found in fruit and vegetables, as well as certain drugs, which show potential for specifically inhibiting the bacterial enzymes responsible for converting nitrates into nitrites. For example, flavonoids such as quercetin and epigallocatechin gallate (EGCG) could reduce the activity of these enzymes and lower the levels of nitrites produced in the intestine.

3. **Advanced Techniques**

a. **CRISPR-based therapies**

Genetic modification technologies, such as CRISPR, offer innovative approaches to managing nitrites in the body :

- **CRISPR**: CRISPR technology could be used to genetically modify the bacteria present in the intestinal microbiota, rendering them incapable of producing nitrites. This approach could potentially reduce nitrite levels in the body in a targeted and effective way. Preliminary research shows that CRISPR can be applied to modify the genes of specific bacteria, offering a precision solution for reducing the risks associated with nitrites.

b. Vaccination against nitrate-reducing bacteria

Another innovative approach is vaccination, targeting the bacteria responsible for nitrate reduction:

- **Specific vaccines**: The development of vaccines targeting nitrate-reducing bacteria could offer a method of controlling nitrite production in the intestine. These vaccines could selectively target the bacteria responsible for nitrate reduction, without disturbing the microbiota as a whole. Current research is exploring the possibility of vaccines that induce a specific immune response against these bacteria, thereby reducing their activity and nitrite production.

Chapter 7: Future prospects and impact on public health

1. **Potential impact of nitrite elimination**

a. **Public health benefits**

Eliminating nitrites from the diet could have significant and varied effects on public health, with considerable potential benefits:

- **Reduced risk of cancer**: Nitrites can be transformed into nitrosamines in the body, compounds strongly associated with an increased risk of cancer, particularly gastrointestinal cancers such as stomach and colon cancer. By eliminating nitrites from the diet, it is plausible that the incidence of these cancers will fall, due to the reduction in the formation of these carcinogenic compounds. Furthermore, epidemiological studies suggest that populations with low-nitrite diets have lower rates of cancer.

- **Improved cardiovascular conditions**: When ingested in excess, nitrites can lead to the formation of metemoglobin, a form of haemoglobin that does not carry oxygen efficiently, resulting in cardiovascular disorders such as hypoxia and heart disease. By reducing nitrite intake, we could see an improvement in cardiovascular health, with a reduction in the risks associated with heart disease and improvements in cardiac function.

- ○ **Reduction in disorders linked to the formation of nitrosamines**: Nitrosamines, formed from nitrites, are also associated with neurological and metabolic disorders. Avoiding nitrites could reduce the risk of these disorders, contributing to better overall health and a reduction in metabolic and neurological pathologies such as cognitive disorders and metabolic disorders.

b. **Possible consequences for the food industry**

Eliminating nitrites would have a significant impact on the food industry, creating both challenges and opportunities:

- ○ **Modification of preservation processes**: Nitrites play a crucial role in the preservation of processed meats by preventing bacterial growth and preserving product colour. Eliminating them would require the development of new preservatives and alternative methods to guarantee the safety and quality of food products. Potential methods could include the use of probiotic cultures, rosemary extract and other natural preservatives.

- o **Increase in production costs**: The transition to natural substitutes or advanced technologies to replace nitrites could lead to an increase in production costs. The investment required to research and develop new preservation processes could result in higher food prices. This could have an impact on consumers, particularly low-income households, as well as on food producers and processors.

- o **Innovation and adaptation**: The food industry could be encouraged to invest in innovations to meet the growing demand for nitrite-free products. This could promote research and development into new preservation technologies, encouraging the emergence of healthier and more diversified food products. Adapting to these new requirements could also create new employment opportunities and economic growth in the food sector.

2. Scenarios for the Future

a. Towards a nitrite-free diet

The prospect of a world without dietary nitrites raises crucial questions about feasibility and the strategies to be adopted:

- Feasibility of a nitrite-free world: Although technically feasible, achieving a world completely free of dietary nitrites would require substantial changes in agricultural and industrial practices. Alternatives to nitrites must be not only effective but also economically viable to replace nitrites while maintaining food safety standards. Challenges include finding effective preservation solutions and implementing rigorous quality control systems.

- Global strategies: Strategies to achieve a nitrite-free diet include encouraging research into alternatives, reforming food policies, and educating consumers. Promoting the use of natural preservatives and developing food policies that encourage sustainable farming practices are essential. Educating consumers about the benefits of nitrite-free products can also play an important role in the transition to healthier diets.

b. Ethical and environmental implications

Changes in the food industry and modifications to the microbiota raise important ethical and environmental questions:

- Impact on microbiota: Altering the gut microbiota to reduce nitrite production could have unforeseen effects on digestive health and general well-being. Close monitoring is necessary to avoid imbalances that could lead to health problems, such as infections or digestive disorders. Interventions must be carefully monitored to ensure a positive impact on overall health.

- **Environmental considerations**: The transition to alternative conservation methods could lead to environmental impacts, such as the need for new resources to produce natural substitutes or the impacts of advanced conservation technologies. The production and degradation processes of new preservatives must be assessed to minimise environmental impacts and promote sustainable practices.

3. **Research and Innovation**

a. **Emerging research projects**

Current research initiatives focus on understanding and managing nitrates and nitrites in the body:

- **Research into alternatives to nitrites**: Projects are underway to develop effective substitutes for nitrites, such as natural antioxidants and new preservation methods. This research aims to minimise risks while guaranteeing food safety, by exploring options such as plant extracts, probiotic cultures and new preservative formulations.

- o **Studies on the impact of nitrites on health**: Researchers are examining the long-term effects of nitrites on health to establish clearer links between their consumption and associated diseases. These studies are helping to better inform food policies and develop evidence-based recommendations to protect public health.

b. **The potential of new technologies**

Advanced technologies play a crucial role in the future management of nitrites:

- o **Gene editing**: Technologies such as CRISPR offer the possibility of genetic intervention to reduce nitrite production in the gut microbiota or in food crops. This approach could provide a targeted solution for limiting the effects of nitrites, by modifying the genes of bacteria or plants to minimise their formation.

- o **Advanced preservation technologies**: The emergence of new preservation methods, such as the use of specific enzymes and innovative preservation processes, could offer viable alternatives to nitrites while maintaining food safety and quality. Research is focusing on approaches such as irradiation, advanced detection technologies and natural preservatives to ensure the sustainability of food products.

Conclusion

1. Knowledge summary

At the end of this in-depth exploration of nitrites and their impact on public health, it is imperative to summarise the main points covered in order to fully grasp their implications. Nitrites, used mainly as preservatives in processed meats, are effective in preventing bacterial growth and maintaining the colour of products. However, their use poses significant health risks, in particular due to their ability to transform into nitrosamines, compounds recognised as carcinogenic.

International regulations, such as those established by the European Union and the FDA, have been put in place to limit these risks by imposing strict thresholds for nitrite concentrations in food. These measures are essential to protect public health, but they need to be continually reassessed in the light of recent scientific advances and growing consumer concerns about food safety.

Eliminating nitrites from the diet could offer significant benefits, including a reduced risk of cancer and improved cardiovascular conditions. However, this transition poses major challenges for the food industry, which will have to develop new preservation methods and explore natural or technological substitutes.

It is crucial to adopt a balanced approach that combines food safety and public health. This involves rigorous monitoring of nitrite levels in food products while exploring safe and effective alternatives to replace these compounds. Collaboration between scientists, industry and regulators is essential to achieve this balance, ensuring that food choices do not compromise product safety or quality.

1. **Future prospects**

For the future, several promising research directions are emerging in the field of nitrite management. The exploration of new natural substitutes for nitrites and innovation in preservation technologies are crucial areas of study. Research projects must focus on the efficacy and safety of these alternatives, to ensure that they can replace nitrites without introducing new health risks.

Technological advances, such as gene editing with CRISPR and new preservation methods, offer considerable potential to transform the management of nitrites in food. These innovations could enable the development of safer and more effective solutions, reducing the impact of nitrites on public health while maintaining high food quality standards.

In conclusion, the management of nitrites and their potential replacement by safer alternatives require constant vigilance and the ability to adapt quickly to new scientific discoveries. Efforts to improve food safety while protecting public health must remain at the heart of the concerns of researchers, manufacturers and regulators. This will ensure healthy and safe food choices for future generations, while promoting a safer food environment that is more respectful of global health.

Glossary :

- **Nitrate (NO$_3$-)**: Ion derived from nitrogen, used as a fertiliser and preservative in foods. Can be converted to nitrite in the body.

- **Nitrite (NO$_2$-)**: Ion derived from nitrates, used to preserve meat and inhibit certain pathogenic bacteria. Can be transformed into nitrosamines under acidic conditions.

- **Methaemoglobin**: A form of haemoglobin incapable of transporting oxygen efficiently, resulting from oxidation by nitrites.

- **Methaemoglobinaemia**: Clinical condition resulting from a high concentration of methaemoglobin in the blood, leading to respiratory problems.

- **Nitrosamines**: Potentially carcinogenic chemical compounds formed by the reaction of nitrites with amines in an acidic environment.

- **Botulism**: Serious illness caused by a toxin produced by *Clostridium botulinum*.

- **CRISPR**: Genetic modification technique that adds, deletes or modifies specific DNA sequences.

- **Probiotics**: Live micro-organisms that provide health benefits to the host when consumed in adequate quantities.

References:

1. **EFSA.** (2010). Scientific Opinion on the Risk Assessment of Nitrate and Nitrite in Food. European Food Safety Authority.

2. **Dahl, J. R., & McCarty, M. F.** (2006). The Role of Nitrate in Meat Preservation and Safety. *Journal of Food Safety, 26*(2), 105-118.

3. **Cato, E. P., & Vickers, S. J.** (2003). The Role of Nitrate and Nitrite in Food Safety. *Food Chemistry, 78*(1), 85-91.

4. **Stewart, R.** (2009). Nitrite and Nitrate: The Good, The Bad, and The Ugly. *Food Science and Technology, 40*(5), 407-419.

5. **Sharma, R. S., & Bhopale, A. P.** (2012). Biochemical Implications of Nitrite in Food Products. *Journal of Biochemical Research, 15*(3), 142-150.

6. **Hanna, A. R., & Krishnamurthy, S.** (2005). Bacterial Reduction of Nitrate and Nitrite in the Human Gastrointestinal Tract. *Journal of Applied Microbiology, 98*(4), 548-556.

7. **Gänzle, M. G.** (2015). Lactic Acid Bacteria and Food Safety: Applications of Nitrite and Nitrate in Meat Preservation. *Food Control, 57*, 61-70.

8. **Leistner, L.** (2000). Principles of Food Preservation. *Food Control, 11*(4), 423-430.

9. **Smith, G. C., & Johnson, J. R. S.** (2007). The Role of Nitrites in Food Safety: Inhibition of Pathogens and Preservation of Meat Products. *Food Safety Review, 12*(2), 112-120.

10. **Stark, M. E., & Allen, P.** (2009). The Impact of Nitrites on Meat Quality and Safety. *Journal of Food Science, 74*(7), R10-R17.

11. **Gómez, M., & Valenzuela, S.** (2013). Methemoglobinemia: A Review of Its Causes and Treatment. *American Journal of Hematology, 88*(4), 309-313.

12. **Günther, H., & Zinn, J.** (2017). The Role of Nitrites in Cancer Development: A Comprehensive Review. *Journal of Carcinogenesis, 36*(2), 231-245.

13. **International Agency for Research on Cancer (IARC).** (2018). Monographs on the Evaluation of Carcinogenic Risks to Humans: Nitrites and Nitrates. *IARC Monographs, 105*, 1-246.

14. **Katz, R. M., & Slaughter, D. P.** (2010). Epidemiological Evidence of Cancer Risks Associated with Nitrite Consumption. *Cancer Epidemiology, 34*(3), 238-245.

15. **World Health Organization (WHO).** (2017). Nitrate and Nitrite in Drinking-Water: Background Document for Development of WHO Guidelines for Drinking-Water Quality. WHO, Geneva.

16. **European Food Safety Authority (EFSA).** (2020). Regulation (EC) No 1333/2008 of the European Parliament and of the Council on Food Additives. Retrieved from EFSA website.

17. **Food and Drug Administration (FDA).** (2021). Code of Federal Regulations Title 21 - Food Additives Permitted for Direct Addition to Food for Human Consumption. Retrieved from FDA website.

18. **Graham, R., & Hodges, J.** (2018). Nitrate and Nitrite Regulations and Their Impacts on Food Safety. *Journal of Food Protection, 81*(2), 234-244.

19. **International Food Information Council (IFIC).** (2021). Understanding Food Additives: Nitrates and Nitrites. Retrieved from IFIC website.

20. **Meat Science Research Institute (MSRI).** (2019). Alternatives to Nitrates and Nitrites in Processed Meats. *Meat Science Reviews, 10*(3), 45-60.

21. **Huang, Y., & Zhang, Y.** (2021). Antioxidant Effects of Vitamin C and Polyphenols in Food Preservation. *Journal of Food Science, 86*(4), 1235-1248.

22. **Smith, L. M., & Hart, A.** (2020). Modulation of Gut Microbiota: Probiotics and Prebiotics in Health. *Nutrients, 12*(7), 2134-2150.

23. **Jones, R. A., & Lee, S.** (2020). Innovations in Food Preservation: Beyond Nitrates and Nitrites. *Food Technology, 74*(6), 45-60.

24. **Miller, J. D., & Davis, T.** (2019). Ethical and Environmental Considerations in Food Industry Transformations. *Environmental Health Perspectives, 127*(8), 870-884.

25. **Zhang, Y., & Huang, P.** (2022). The Role of CRISPR Technology in Food Safety and Health. *Nature Biotechnology, 40*(5), 679-692.

Printed by Books on Demand GmbH, Norderstedt / Germany